National Audit Office

AF885305

Department for Communities and Local Government

Local government funding: Assurance to Parliament

Report by the Comptroller and Auditor General

Ordered by the House of Commons
to be printed on 24 June 2014

This report has been prepared under Section 6 of the National Audit Act 1983 for presentation to the House of Commons in accordance with Section 9 of the Act

Amyas Morse
Comptroller and Auditor General
National Audit Office

23 June 2014

HC 174 | £10.00

This report examines how the government funds local authorities, the conditions attached to this funding, and how it gives Parliament assurance on how local authorities use that funding.

10471 06/14 NAO

Contents

Key facts 4

Summary 5

Part One
Funding for local authorities 13

Part Two
Assurance on how local authorities spend grants 21

Part Three
Assurance on spending from the local accountability system 29

Glossary 40

Appendix One
Our audit approach 42

Appendix Two
Our evidence base 44

Appendix Three
Government grants paid to local authorities, 2013-14 46

The National Audit Office study team consisted of:
James Beveridge, Simon Bittlestone, Alex Burfitt, Jordan Fung, Claire Fears, Paul Hirst and John O'Grady, under the direction of Aileen Murphie and Mike Suffield.

This report can be found on the National Audit Office website at www.nao.org.uk

For further information about the National Audit Office please contact:

National Audit Office
Press Office
157–197 Buckingham Palace Road
Victoria
London
SW1W 9SP

Tel: 020 7798 7400

Enquiries: www.nao.org.uk/contact-us

Website: www.nao.org.uk

Twitter: @NAOorguk

Key facts

£36.1bn

government funding to local authorities in 2013-14 (exc. payments to schools and individuals that pass through local authorities)

8

government departments fund local authorities

61

main grants paid to local authorities, 2013-14

£3.2 billion	in **ringfenced grants** to support local authorities to deliver their statutory duties – has specific conditions on how local authorities can spend it
£25 billion	in **unringfenced general grants** to support local authorities to deliver their statutory duties – no conditions on how local authorities spend it
£7.8 billion	in **unringfenced targeted grants** to support local authorities to deliver their statutory duties – where departments expect, but cannot require, that local authorities spend funding on a specific activity
100 per cent	of **ringfenced funding** for which local authorities report on how they have used the funding
0 per cent	of **unringfenced general funding** for which local authorities report on how they have used the funding
64 per cent	of **unringfenced targeted funding** for which local authorities report in varying degrees of detail, on how they have used funding

Summary

1 Government departments must assure Parliament that their spending meets Parliament's intentions and is value for money. For departments that fund local authorities, they must assure themselves that funding they give local authorities meets the same Parliamentary requirements. Since 2010-11, the nature of departments' information to secure assurance on local authority funding has changed, and the volume has significantly decreased.

2 The government set out changes in the 2010 spending review to implement policy to give local authorities more control over their funding, letting them allocate resources to meet local priorities. The government also wanted to reduce the reporting burden for local authorities. The government intended to provide local authorities with flexibility to fulfil their statutory duties despite funding reductions. These changes accelerated the trend of previous governments to increase local authorities' financial flexibility by reducing the number and value of ringfenced grants.

3 The government has made these changes by doing the following:

- **Removing direct conditions (ringfences) and reporting mechanisms from individual grants.** It has also combined grants for specific purposes into larger payments, which local authorities can spend across many activities.
- **Withdrawing some of the detailed frameworks for monitoring local authority spending and performance.** It relies instead on the overall system of local accountability for assurance. This system consists of checks and balances such as inspectorates, external audit, statutory duties on some local authority officers, and local councillors being democratically accountable. Local authorities also provide data to departments on their spending and performance, via a number of data sets contained in the single data list.

4 The new arrangements provide departments with less direct information on how local authorities spend government grants and whether they achieve their outcomes. Departments rely more on systems of local accountability to ensure relevant local authority spending meets with Parliament's intentions and is value for money.

5 The Department for Communities and Local Government (the Department) led the government's changes to local government funding. It is the department primarily responsible for the new assurance arrangements. The Department has led the reduction in the number of grants and removed restrictions on how local authorities can spend them. It coordinates the design of new grants to local authorities from across government. Other departments increasingly channel funding to local authorities through the Department. This is because of the policy of combining separate grants into larger grant payments for multiple service areas.

Our report

6 This report examines how the Department has implemented and oversees the assurance framework that enables departments to assure Parliament on funding for local authorities:

- Part One examines how the departments funded local authorities in 2013-14, how that changed since 2010-11, and how the Department made these changes.
- Part Two examines whether departments' monitoring of local authorities gives them enough assurance on whether local authority grants are used in line with Parliament's intentions and sufficient information on the impact of the grants on policy objectives.
- Part Three evaluates whether the Department ensures the local accountability system is effective for providing assurance on value for money, and considers emerging issues and risks.

7 The report complements our report on the Department's 2013-14 accounts. In that report, we describe how we have carried out our financial audit responsibilities in the context of the Department's current operating environment.

8 Local authorities receive several different streams of funding from government. In 2013-14 departments passed £53.3 billion through local authorities to support schools and pupils or to fund welfare and training payments for individuals. Local authorities have almost no influence over the use of this funding. We do not examine this funding in this report.

Key findings

Local authority funding and assurance, 2013-14

9 **In 2013-14, the government gave local authorities £36.1 billion in funding (excluding funding passed directly to schools and individuals).** Revenue funding accounted for £32.0 billion, with £4.1 billion in capital funding. Eight different departments gave funding. The Department, which coordinates funding to local authorities, gave £26.5 billion (73 per cent). This funding complements a range of non-departmental sources of funding used by local authorities such as council tax receipts, income from sales, fees and charges and capital receipts (paragraphs 1.3 to 1.6).

10 **In order to support the delivery of their core functions and statutory duties, the government funds local authorities through several grant types, which have varying conditions attached.** These include:

- **Ringfenced grants**
  Local authorities must spend this on a specific activity, for which they give the government evidence through rigorous reporting.
- **Unringfenced general grants**
  There is no expectation of how local authorities spend this funding, other than they should spend it lawfully.
- **Unringfenced targeted grants**
  Departments provide these grants for local authorities to fund specific activities. However, as they are not ringfenced, local authorities can reallocate these grants to other areas of their work to meet local priorities (paragraphs 1.7 to 1.13).

11 **The government gives local authorities most funding via unringfenced grants, to maximise local flexibility.** In 2013-14, £25 billion (69 per cent) of government funding to local authorities was through unringfenced general grants. £7.8 billion (22 per cent) was paid in unringfenced targeted grants. In both cases, there are no conditions requiring local authorities to use funding for a specific purpose (paragraphs 1.9 to 1.10).

12 **The Department has reduced conditions on how local authorities must use grants and merged some grants since the 2010 spending review.** Following the spending review, the Department led an effective review of whether ringfences should remain and whether individual grants should roll into larger grants. It communicated clear guidance to other departments that grants should be ringfenced only in very exceptional circumstances. The Department undertook a second round of rolling individual grants into larger ones as part of the introduction of the Business Rates Retention scheme in 2013-14, which was a significant change to how the government funds local authorities (paragraphs 1.15 to 1.23).

13 The Department has made changes since 2010 that have increased the share of funding that has no specific intentions, other than to support local authorities deliver their statutory duties and core functions. Like-for-like (adjusting for recent changes in local government duties as far as possible) ringfenced revenue funding fell from 7 per cent to less than 1 per cent. The level of unringfenced general revenue funding, where government has expressed no intention for its use, increased from 67 per cent of local authority funding in 2010-11, to 84 per cent in 2013-14. Unringfenced targeted funding made up 15 per cent of local authorities' government revenue funding in 2013-14, compared with 26 per cent in 2010-11 (paragraphs 1.24 to 1.27).

Assurance on how local authorities spend government grants

14 All ringfenced grants have clear reporting mechanisms, where local authorities must show that they have met grant conditions. The government monitors how local authorities use these grants for assurance that local authority expenditure is in line with Parliament's intentions (paragraphs 2.15 to 2.16).

15 In contrast, local authorities do not have to show how they spend unringfenced general grants; none of these grants require local authorities to report back directly to departments. These grants, which include Revenue Support Grant and Council Tax Freeze Grant, are for general rather than specific purposes. The government puts no conditions on this funding. For regularity, departments only need to assure Parliament that local authorities have used the funding within their legal powers. Equally, as there are no specific stated policy objectives for how local authorities should use these grants, departments do not monitor spending patterns directly to assess policy impact. The local accountability system gives assurance on whether this funding is lawful and value for money. Unringfenced general grants form most local authority funding from government. This means that direct monitoring of how local authorities use funding overall is limited. These changes are a specific policy intent of localism (paragraph 2.17).

16 Departments do not have to monitor directly how local authorities use unringfenced targeted grants. However, departments often monitor their use. The Department tries to manage the scale and nature of this monitoring, as part of the government's policy to reduce the reporting burden on local authorities. The Department has given other departments guidance on the nature of information they can collect. It encourages them to monitor these grants through existing data sets rather than new ones. The Department also encourages other departments to consider other factors that affect how local authorities use funding, when designing reporting mechanisms. These include whether the grant is for a statutory purpose and whether the local authority secured the funding through competitive bidding. The Department considers that statutory duties, or having business plans and bid documents, shape local authorities' spending decisions and therefore reduce the need for direct reporting (paragraphs 2.18 to 2.21).

17 **The policy of reducing reporting burdens means that for £2.8 billion of the £7.8 billion in unringfenced targeted grants, departments receive no direct information on how local authorities use them.** The lack of direct evidence means that departments cannot be confident of the impact of these grants and whether they should continue, change or end them. However, for most (£2.3 billion) of unringfenced targeted grants where there is no direct reporting, there are other factors in place such as statutory responsibilities, business cases or memorandums of understanding. The Department feels that this gives other departments assurance that local authorities will use the funding in line with the policy intention (paragraphs 2.22 to 2.29).

Departments' assurance over value for money locally

18 **Central government relies on the system of local accountability for assurance over the value for money of funding it gives local authorities.** The system includes local checks and balances. These include the activities of external auditors, having local authority officers with legal duties, and having democratically accountable local councillors. The Department describes this system in its accountability system statement. This includes mechanisms that relate specifically to assurance on value for money. Other departments secure assurance for the funding they give to local authorities from the core accountability system set out in the Department's system statement (paragraphs 3.3 to 3.8).

19 **The Department is clear that its role is to assure itself that the local accountability system is effective, rather than to monitor whether spending is value for money.** The Department's core principles state that local authorities' prime accountability is to their local electorate, and that local councillors are best placed to decide what is value for money locally. Consequently, the Department thinks that local authorities will achieve value for money differently, according to local priorities. The Department gets assurance over the value for money of government grants to local authorities from the existence of the checks and balances in the local accountability system, which are intended to promote sound and informed local authority decision-making (paragraphs 3.9 to 3.11).

20 **The Department has improved how it assures itself that the local accountability system is effective.** In September 2013, the Department updated its system statement to include the information it uses to oversee the local accountability system. The Department also advised the Accounting Officer on the system's effectiveness in December 2013. It concluded that the system did not require any changes. The Department will continue to advise the Accounting Officer twice a year. The Department is considering how to address assurance issues that result from the closure of the Audit Commission given the Commission's role in collating information provided by auditors (paragraphs 3.16 to 3.20).

21 **Despite recent improvements, it is not clear that the Department knows whether the system is effective in securing value for money.** The Department believes that the system creates the conditions for local authorities to achieve value for money through pressure to improve outcomes, reduced incomes and greater transparency of their spending decisions. However, the Department's monitoring information gives limited insight into whether this is happening in practice. Instead, its monitoring focuses more on financial and service sustainability. The Department's advice to the Accounting Officer in December 2013 did not consider whether the system is effective in relation to value for money (paragraphs 3.21 to 3.22).

22 **There is a tension between departments using unringfenced targeted grants and relying on the local accountability system.** Whilst these grants lack formal conditions, departments seek to exert influence over local authorities' use of the funding by establishing spending expectations for the grant, rather than leaving the decision solely for local consideration. In the Department's view, these types of grants allow national priorities to be pursued locally while also providing a degree of local financial flexibility. For example, local authorities can reallocate unspent grant funding to other activities rather than repaying it to departments. However, the primacy of local priorities within the accountability system could mean that the departments' expectations for these grants are overridden locally. Overall, departments wanting to achieve specific objectives through their funding to local authorities is not fully consistent with an assurance and accountability framework designed to promote local priorities, and where there is limited reporting on local spending (paragraphs 3.24 to 3.29).

23 **Public funding comes increasingly through multi-agency, cross-border organisations, which do not fit easily with government relying on the local accountability system.** For example, Local Enterprise Partnerships, run by local businesses and local councillors, can have members from many local authorities, with one local authority overseeing how funding is distributed and used. One local authority could therefore allocate funds on behalf of other local authorities, subject to the parameters of a partnership agreement. This blurs the lines of accountability between local authorities and their electorates, on which the local accountability system depends. The statutory duties of local authority officers become more complex, as they apply only to a single local authority. Though they retain responsibility for allocating funds to the Partnership, they may not be involved ultimately in decisions over how it is spent. The growth of these sorts of arrangements will place increasing pressure on the core principles of the Department's local accountability system as it is currently designed (paragraphs 3.30 to 3.31).

24 **A system of accountability in which local authorities and other local public bodies report to individual departments is at odds with emerging patterns of local service delivery in which local bodies from different sectors pool budgets and work across institutional boundaries to tackle complex local issues.** New arrangements, such as Health and Wellbeing Boards, pooled health budgets, and multi-agency initiatives to tackle complex local social problems, mean that local authorities are increasingly designing and delivering services jointly with health bodies and other public sector agencies locally. Despite this local integration, different public bodies remain accountable to different departments via separate accountability systems. The Department needs to understand the fit between the different departmental accountability systems to ensure that jointly delivered local schemes are suitably accountable to the relevant departments, whilst ensuring that these accountability arrangements are not detrimental to joint local working (paragraphs 3.31 to 3.33).

Conclusion

25 The Department has implemented the government policies to increase local authorities' flexibility over their funding and reduce reporting burdens. Local authorities have more freedom to allocate resources according to local priorities. However, at the same time, reductions to government funding have put increasing pressure on budgets. The Department's arrangements to assure Parliament over funding are in transition as government adapts to receiving less evidence on how local authorities spend funding. It is moving to a model of local accountability focused on understanding, rather than shaping, how local authorities use funding. Tensions remain where departments continue to specify policy objectives for grants, despite local priorities now taking precedence for how funding can be spent.

26 Under the new arrangements the Department does not have to monitor local spending decisions. However, it still must assure itself that the local accountability system that oversees local spending is effective. The Department has improved its understanding of how the system prevents financial and service failure. It could do more to understand whether the system is effective in delivering value for money.

Actions to consider

27 It is for the Department to set out, and Parliament to consider, current assurance arrangements as set out in our report. Our findings, though, lead us to suggest a number of actions departments should consider concerning their assurance over funding granted to local authorities:

a Assess the appropriateness of continuing to fund local authorities through targeted grants in the context of an approach to value for money based on local priorities.

b Departments may judge that unringfenced targeted grants are appropriate. If so, they should assess how far reporting arrangements for targeted grants give enough assurance that local authorities spend this funding according to policy intentions.

c Consider value for money as well as financial and service sustainability, when assessing whether the local accountability system is effective.

d Consider updating the guidance on specific grant determinations, as it is now outdated.

e Review their accountability system statements against ongoing changes to public services involving partnerships and cross-border working.

Part One

Funding for local authorities

1.1 Since 2010-11, local authority funding has changed substantially. Departments have worked to implement the government's policies of providing greater financial freedoms and flexibilities to local authorities and reduce reporting burdens. To do so, they have removed ringfences and combined some grants. The overall scale of funding that departments give local authorities has also fallen, to meet spending reductions set out in the 2010 spending review.[1]

1.2 This part of the report sets out the scale, structure and sources of local authority funding and examines how these have changed since 2010-11.

How local authorities are funded

Sources of local authority funding

1.3 Local authorities[2] receive funding from several sources.[3] In 2012-13,[4] non-departmental sources such as Council Tax, fees and charges, and interest and investments accounted for £33.8 billion (29 per cent) of their revenue income. Income from capital receipts and non-government grants and contributions accounted for £2.5 billion (31 per cent) of capital income. Overall, however, funding voted by Parliament and channelled through eight different departments accounted for most local authority income (71 per cent) (**Figure 1** overleaf).[5]

1.4 However, there is marked variation at the local level in the significance of departmental funding as a source of local authority income. Among the 150 single tier and county councils,[6] the largest authorities with responsibility for education and social care, there are seven local authorities where government funding represented less than 60 per cent of their income in 2012-13, and 14 where it accounted for more than 80 per cent.

1 HM Treasury, *Spending Review 2010*, October 2010, available at: www.gov.uk/government/uploads/system/uploads/attachment_data/file/203826/Spending_review_2010.pdf

2 We use 'local authorities' to mean the 353 councils in England. Other types of local authority including National Park Authorities, Police and Crime Commissioners, Fire and Rescue Authorities and the Greater London Authority, are excluded.

3 We exclude income from local authorities' Housing Revenue Accounts from this analysis. Housing Revenue Accounts are where local authorities account for all expenditure incurred and income received in respect of the provision of rented accommodation.

4 Data on local authority income from local sources is only available for 2012-13, so data on total local authority income is shown for this year. Throughout the rest of the report where we focus on government funding only, data for 2013-14 is used.

5 This includes funding raised via Business Rates. These are collected by local authorities and returned to government for redistribution back to local authorities. Since 2013-14, local authorities have retained a share of these locally.

6 City of London and the Isles of Scilly are excluded.

Figure 1
Sources of local authority income in 2012-13

Local authorities receive funding from a number of sources other than central government

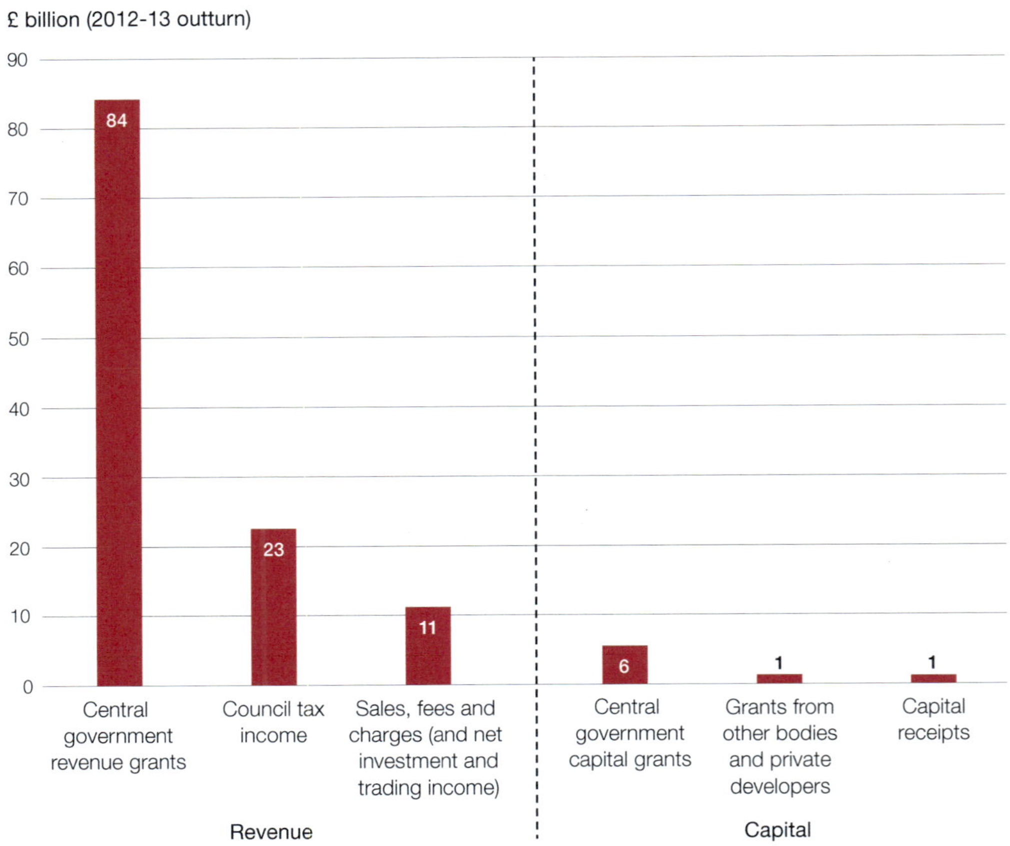

Source: National Audit Office analysis of Department for Communities and Local Government's data on sources of local authority income – Revenue Outturn and Capital Outturn Return data

Central government funding to local authorities

1.5 In 2013-14, local authorities in England budgeted to receive £89.4 billion in income from the government. This was made up of £85.1 billion in revenue funding and £4.2 billion in capital funding. This funding falls into three distinct groups, which have significant implications for how local authorities can use the funding:

- £31.2 billion in frontline funding for schools from the Department for Education including Dedicated Schools Grant (£29.2 billion), Pupil Premium Grant (£1.8 billion), and Devolved Formula Capital Funding (£128 million). This funding is ringfenced and local authorities transfer it directly to schools.[7]
- £22 billion in revenue 'grants outside external aggregate finance', which are grants that local authorities pass directly to individuals, as Housing Benefit or training support payments.
- £36.1 billion in revenue and capital grants that local authorities retain in order to support the delivery of their statutory functions and duties. Eight different departments gave funding. The Department, which coordinates funding to local authorities, gave £26.5 billion (73 per cent).

1.6 Local authorities do not control the first two funding streams, so our analysis excludes these and instead we focus on their core controllable funding (£36.1 billion). This reflects the Department for Communities and Local Government's (the Department's) approach in its accountability system statement.[8] Departments gain assurance on the first two funding streams through separate arrangements run by the relevant departments, principally the Departments for Education and Work & Pensions.

Types of grants

1.7 Departments give local authorities £36.1 billion through several types of grants to support the delivery of their core functions and statutory duties. The type of grant used affects the degree of control local authorities have over the funding, and the assurance departments expect over its use (**Figure 2** overleaf).[9]

1.8 Our analysis of the grants shows that departments gave local authorities **£3.2 billion in 2013-14 through 10 ringfenced grants** (Appendix Three). Local authorities must use this funding for specified purposes. Two ringfenced grants, Public Health Grant (£2.7 billion) and Local Major Transport Schemes (£456 million), accounted for 96 per cent of all ringfenced funding.

7 Department for Communities and Local Government, *Accounting Officer Accountability System Statements for Local Government and for Fire and Rescue Authorities*, September 2013, p.5.
8 Department for Communities and Local Government, *Accounting Officer Accountability System Statements for Local Government and for Fire and Rescue Authorities*, September 2013.
9 All capital grants are provided under the condition that they are used solely for capital purposes.

Figure 2

Types of government grant, 2013-14

Funding type	Amount 2013-14 (£bn)	Grant conditions	Example
Ringfenced grants	3.2	Local authorities must use the funding for a specific purpose. Departments can claw back funding if local authorities do not use it, or do not spend it in line with funding conditions	Public Health Grant
Unringfenced general grants	25.0	None. Local authorities can use this funding for any purpose within their legal powers	Revenue Support Grant
Unringfenced targeted grants	7.8	None. These are unringfenced grants but departments intend for local authorities to use this funding for a specific purpose. Nonetheless local authorities can reallocate funding to other activities should they choose, and departments typically have no power to claw back this funding	Local Sustainable Transport Fund

Source: National Audit Office analysis of Department for Communities and Local Government's data

1.9 Unringfenced grants, where departments do not place specific conditions on how local authorities can use the funding, fall into two groups. The largest group are unringfenced general grants for local authorities to provide their core functions and statutory duties. There are no conditions or expectations on how local authorities use the funding, other than that the spending is lawful. **In 2013-14, departments gave local authorities £25 billion through six unringfenced general grants** (Appendix Three). The largest components were Revenue Support Grant (£14.5 billion) and retained income from the Business Rates Retention scheme (£9.5 billion). Others are Council Tax Freeze Grant (£265 million) and New Homes Bonus (£668 million).

1.10 Unringfenced targeted grants, which the Department calls 'non-ringfenced specific grants', are a second, smaller group of unringfenced grants. In 2013-14, departments gave local authorities £7.8 billion through 45 separate grants of this type (Appendix Three). Departments generally provide these grants under Section 31 of the 2003 Local Government Act, which gives ministers powers to make direct grants to local authorities.

1.11 As with unringfenced general grants, departments do not attach conditions to unringfenced targeted grants that require local authorities to spend them on something specific. However, unlike unringfenced general grants, departments expect local authorities to use unringfenced targeted grants for specific purposes. This is set out in each grant's determination, which grant-giving departments give local authorities explaining the amount and purpose of each grant. Ultimately, however, there are no conditions attached to this funding. Local authorities can allocate the funding how they choose, and departments have no powers to claw funding back.

1.12 Sometimes departments provide funding in unringfenced targeted grants, rather than unringfenced general grants, in order to make funding decisions transparent to local authorities. For example, the Department for Education provides the Education Services Grant separately from general formula grant funding. The Department for Education designed the grant to be transparent, giving local authorities a clear understanding of how they calculated their allocation in the context of local authority maintained schools converting to academies.

1.13 All capital grants fall into the ringfenced or unringfenced targeted grants categories.[10] None are classified as general unringfenced funding because all have some specification. Departments expect this funding to be used for certain purposes, such as for schools or highways maintenance.

1.14 Local authorities' prudential borrowing powers represent the capital equivalent of general unringfenced funding.[11] Local authorities borrowed £6 billion this way in 2013-14, which gives them total control over how they can use the funding. Departments do not provide this funding so we have not analysed it.

Changes in funding to local authorities

Government's intention to increase local flexibility

1.15 Our analysis indicates that 91 per cent of departmental funding to councils is unringfenced, with 69 per cent via unringfenced general grants and 22 per cent unringfenced targeted grants.

1.16 The structure of local government funding and the lack of grant conditions reflects government policy to increase local financial flexibility. In the 2010 spending review, government said it wanted to increase local authorities' flexibility over how they can spend their funding. The government wanted local authorities to be able to respond to local priorities, rather than being directed by conditions that central government places on spending.

10 All capital grants, whether they are ringfenced or not, are subject to the condition that local authorities must use the funding for capital purposes.

11 Prudential borrowing is unsupported borrowing funded by local authorities' own resources, such as Council Tax, rather than through government grants.

1.17 The main changes the government proposed were:

- removing ringfences from grants;
- reducing the number of separate grants paid to local authorities, to simplify funding; and
- shifting accountability for how funding is spent to local authorities, and away from central government.

1.18 The government linked its policy to increase local flexibility with its reduction in funding to local authorities after the 2010 spending review. The review set out the government's plans to reduce its funding of local authorities by 26 per cent over four years.[12] The government wanted to give local authorities freedom to be innovative and efficient, against increasing financial challenges.

1.19 The 2011 Localism Act reinforced the government's intentions to devolve financial decision-making. The Act gave local authorities a "general power of competence". Local authorities can now act in any way that does not break other laws.

1.20 Changes in local authority funding since 2010-11 continue a process begun in the 2007 comprehensive spending review. This saw the creation of Area Based Grant, which, by 2010-11, contained 55 different funding streams worth £4.6 billion. Many of these were previously ringfenced. To meet departmental objectives for this funding, the government set up a rigorous framework to monitor local outcomes using 'local area agreements', an expanded 'national indicator set' and the Audit Commission's 'comprehensive area assessments'. The government has abolished or scaled back these local monitoring arrangements since 2010-11.

The Department's role

1.21 The Department has led changes to local authority funding since the 2010 spending review. It gave all government departments guidance, which said that they should first look to roll their grants into the Department's general unringfenced formula grant for local authorities. Failing this, departments should give targeted unringfenced grants. If neither of these options were appropriate, then departments could provide ringfenced grants.

12 Comptroller and Auditor General, *Department for Communities and Local Government: Financial sustainability of local authorities*, Session 2012-13, HC 888, National Audit Office, January 2013.

1.22 The Department challenged other departments to justify any conditions on their grants to local authorities. Combined with funding reductions, this brought significant changes in the structure of revenue funding, including:

- Forty-one revenue funding streams, worth £1.4 billion, many of which had been part of the Area Based Grant, were ended;
- Eighteen revenue funding streams from the Department for Education previously worth £2.9 billion, of which five had been ringfenced, were rolled into a new unringfenced Early Intervention Grant; and
- Twenty-one revenue grants previously worth £3.2 billion were rolled into general unringfenced funding.

1.23 The Department undertook a second round of restructuring of revenue funding before introducing the Business Rates Retention scheme in 2013-14. This saw the Early Intervention Grant (£1.1 billion)[13] and the Learning Disability and Health Reform Grant (£1.4 billion) rolled into general unringfenced funding. Localising Council Tax support also increased general unringfenced funding for local authorities by £3.1 billion.[14] Under the new arrangements, local authorities receive a fixed amount of funding and administer their own Council Tax support scheme.[15]

1.24 The Department has maintained its approach throughout this period and has to sign off all new Section 31 grants that departments intend to give to local authorities. HM Treasury also ensures that departments do not attach ringfences to new grants.

1.25 Since 2010-11, the government has introduced new responsibilities for local authorities, such as for Public Health and Council Tax support. We have adjusted the data to estimate a like-for-like comparison between funding in 2010-11 and 2013-14.[16] There is clear evidence of a shift away from ringfenced and targeted funding, and an increase in unringfenced general funding (see **Figure 3** overleaf). Like-for-like, ringfenced funding fell from 7 per cent to less than 1 per cent. The level of unringfenced general funding increased from 67 per cent of local authority funding in 2010-11, to 84 per cent in 2013-14. Unringfenced targeted funding fell from 26 per cent of government revenue funding for local authorities in 2010-11, to 15 per cent in 2013-14.

13 At the same time as this transfer to general unringfenced funding, £520 million of the Early Intervention Grant was moved into the Dedicated Schools Grant, another £150 million was used to establish the Adoption Reform Grant.

14 Funding for Council Tax support had previously been paid 'outside aggregate external finance', meaning it passed through local authorities to recipients via a grant.

15 Comptroller and Auditor General, *Council Tax support*, Session 2013-14, HC 882, National Audit Office, December 2013.

16 This is an estimate that adjusts for the major changes between the two years in order to provide an illustration of the pattern of change in types of grant. It is not a definitive analysis of the change in overall government funding to local authorities on a like-for-like basis over this period.

Figure 3

Change in budgeted revenue funding by grant type, 2010-11[1] to 2013-14[2]

On a like-for-like basis, the level of ringfenced and targeted funding has fallen since 2010-11 while unringfenced general funding has grown

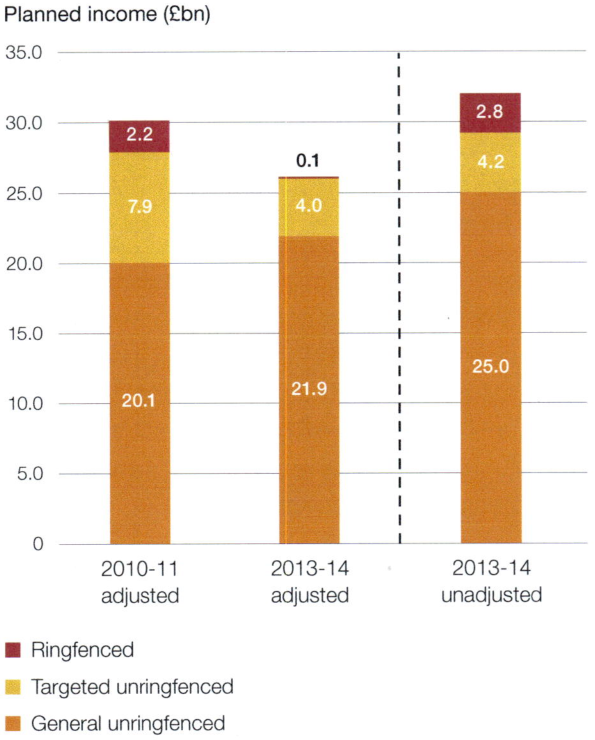

Notes

1 2010-11 data is before the 2010 emergency budget and the 2010-11 in-year cuts.

2 The adjusted data for 2010-11 includes learning disability commissioning transfers, and excludes two-year-old early learning and childcare funding. The adjusted data for 2013-14 excludes Public Health grant, local welfare provision and Council Tax support, new burdens and transition funding.

Source: National Audit Office analysis of Department for Communities and Local Government's data

1.26 Comparisons for capital funding over this period are complicated because the Department for Education changed its capital programme, including cancelling Building Schools for the Future. The continued growth in the number of Academies, and the introduction of Free Schools, both of which the Department for Education funds directly, has also reduced the amount of capital schools funding it pays local authorities. Funding from the Department for Education to local authorities changed from a planned budget of £5.3 billion in 2010-11, nearly all of which (94 per cent) was ringfenced, to £1.6 billion in 2013-14, none of which was. This reduction was as a result of the current government's changed spending plans.

1.27 Capital funding from other departments to local authorities changed from £2.4 billion in 2010-11 (56 per cent ringfenced) to £2.5 billion in 2013-14 (18 per cent ringfenced).

Part Two

Assurance on how local authorities spend grants

2.1 Departments gain assurance on local authority spending of government grants through two main information mechanisms:

- **Direct assurance via reporting on certain individual grants**
  Local authorities report spending, and occasionally outcomes, back to the relevant departments as part of the grant conditions. In these cases, departments (and Parliament) can 'follow the money'.
- **Indirect assurance via the local accountability system**
  Local checks and balances assure that local authorities have used funding lawfully and with the best value for money locally.

2.2 This section examines current arrangements for direct assurance. Part Three then examines indirect arrangements for assurance.

Purpose of direct assurance

Assurance on regularity

2.3 Accounting Officers must assure Parliament about the regularity of their spending. There are two aspects to this:

- Departmental spending must align with Parliament's intentions. These are in each 'ambit', which is part of each department's Main Estimate of their anticipated annual spend. Parliament votes on each estimate, which allows departments to spend funding in line with the programme and the ambit in the estimate. In general, ambits are broad and give departments significant flexibility.
- Departmental spending must be "compliant with the relevant legislation (including EU legislation), [and] delegated authorities".[17] To spend or distribute funding, departments may need delegated authority from the Treasury. Departments must also ensure that there is appropriate legislation, to let them spend the funding as they intend.

17 HM Treasury, *Managing Public Money*, July 2013, Box 2.4.

2.4 Direct assurance is where a department receives precise information on how a local authority has used funding. This is the main way departments get assurance on regularity. It shows whether local authorities have used funding in line with a department's ambit and any legal conditions, and therefore whether spending is regular.

2.5 In principle, direct assurance is only required for ringfenced grants. These are the only grants where local authorities must use funding in a particular way. If a local authority uses a grant 'irregularly', departments may look to claw back the grant.

2.6 Unringfenced general grants have no specific objective other than to help the local authorities provide their functions and duties. Departments provide this funding under legislation that does not automatically impose conditions on grant usage. Consequently, departments do not require direct evidence of how local authorities have used these grants. Accounting Officers can meet their assurance obligations to Parliament without this direct evidence. Departments do still require assurance that local authorities have used this funding lawfully, however. Departments achieve this through other aspects of the local accountability system – principally the work of the local auditor[18] and each local authority's Section 151 officer.

2.7 Departments have specific objectives for unringfenced targeted grants. But these grants are unringfenced, so local authorities do not have to spend them on specific things. Departments do not have to monitor spending to give Parliament assurance on it. The local accountability system gives assurance on whether local authorities have used these grants lawfully.

Assurance on policy impact and effectiveness

2.8 For regularity, departments only need to assure Parliament that local authorities have used the unringfenced targeted grants for any activity within their legal powers. Departments are content that local authorities can spend this funding flexibly. However, departments often monitor this spending to assess the impact, effectiveness and value for money of their spending. This reflects the principles set out in *Managing Public Money*, which says departments should give ministers "enough evidence about the impact of their policies to decide whether to continue, modify or end them".[19]

18 Local auditors do not have a duty to provide assurance on individual spending programmes as part of the audit of general purpose financial statements. However, they are expected to report any material concerns as part of their audit opinion.

19 HM Treasury, *Managing Public Money*, July 2013, paragraph 1.5.3.

Policy changes on direct reporting

2.9 Since removing ringfences, and introducing powers of competence for local authorities in the 2011 Localism Act, the risk of local authorities not using government funding in line with Parliament's intentions is lower. The burden for direct reporting on how local authorities spend funding has also been reduced substantially.

2.10 The government's policy to reduce the reporting burden on local authorities has also changed how much information departments have to draw on to assess the impacts and effectiveness of their spending. To implement the government's policy of increasing local flexibility the Department for Communities and Local Government (the Department) controls how far departments monitor unringfenced targeted grants by other departments. Departmental guidance is clear that grants to local authorities should not have "any requirement to provide monitoring information over and above the National Indicator set".[20]

2.11 The Department encourages other departments to rely on ways other than direct reporting to gain assurance that local authorities have used funding for the specified purpose. These include:

- relying on many of these grants being for statutory purposes – which implies that local authorities face unavoidable costs that the grants meet;
- using existing data returns already in the single data list;[21]
- using memorandums of understanding;
- requiring business cases to access funding, often as part of a competitive bidding process; and
- using staged payments or payment by results.

2.12 In some cases, local authorities give departments no information on how they use grants. Departments instead assume that local authorities spend funding in line with policy objectives due either to the constraints of local authorities' statutory duties, or the expectation that local authorities will honour the commitments in their business cases or memorandums of understanding.

2.13 The lack of direct information on how local authorities have used spending is not a regularity issue, as these grants are unringfenced. However, departments do not have direct evidence of the impact of this spending.

20 Department for Communities and Local Government, *The Preparation of Specific Grant Determinations for Local Authorities: A guide for government departments in England*, July 2011, p.7.

21 The single data list is a list of all the datasets that local authorities must submit to central government.

Patterns and types of direct assurance

2.14 We have used the framework set out in **Figure 4** to show the types of direct assurance associated with 61 different grants provided to local authorities. Of the £36.1 billion provided to local authorities, 78 per cent has no reporting, 21 per cent involves reporting from the grant recipient and 1 per cent involves third-party scrutiny.[22]

Direct reporting on ringfenced grants

2.15 All ringfenced grants have some form of post-award reporting (see Appendix Three). The largest ringfenced revenue grant, Public Health grant (£2.7 billion), which accounts for 96 per cent of ringfenced revenue funding, requires each local authority to publish how they have used the grant each year. The chief executive must certify that the published data accurately reflects expenditure.

2.16 Local Major Transport Schemes funding (£456 million budget in 2013-14), accounts for more than 99 per cent of ringfenced capital funding. It is subject to third-party scrutiny. Local authorities submit quarterly claims and are paid in arrears based on the value of work done. Grant claims are then subject to audit certification by the internal local authority auditor, before undergoing certification by an independent external auditor.

Figure 4
Different types of direct reporting mechanism

Reporting source	Reporting type	Reporting mechanism	Example of grant
No post-award reporting	No reporting or pre-award approval	No reporting	Revenue Support Grant
Reporting from grant recipient	Reporting on activity spend	Data published on total spending for an activity (which may include grant funding)	Highways Maintenance Grant
	Reporting on grant spend	Data published on specific use of grant funding	Public Health Grant
Third-party reporting	Scrutiny from grant funding body	Programme review undertaken by grant funding body	Flood Management
	Independent scrutiny	Programme review undertaken by external body	Right to Control Trailblazers

Note

1 In addition to these reporting mechanisms, local authorities are required to provide assurance that capital grants, whether they are ringfenced or not, have been used for capital purposes.

Source: National Audit Office analysis

22 Levels of the main form of third-party scrutiny, grant certification, have fallen as levels of ringfencing have declined, and also as a result of changes in arrangements for grant certification in preparation for the closure of the Audit Commission, which previously oversaw this activity. Of the £36.1 billion, only the Department for Transport's Local Major Transport Schemes grant is subject to grant certification. A small number of grants outside aggregate external finance, notably Housing Benefit payments, are also subject to grant certification but are outside the scope of this study.

Direct reporting on unringfenced general grants

2.17 All unringfenced general grants (£25 billion) in 2013-14 involved no direct reporting from local authorities on how they had used the funding (see Appendix Three). These grants, which include Revenue Support Grant, locally retained Business Rates and Council Tax Freeze Grant, are for general rather than specific purposes. There are no conditions on this funding. For regularity, departments only need to assure Parliament that local authorities have used the funding for any activity within their legal powers. Equally, as there are no specific stated policy objectives for how local authorities should use these grants, other than to contribute to the delivery of local authorities' core functions and statutory duties, departments do not monitor spending patterns directly to assess policy impact. The local accountability system gives assurance on whether this funding is used lawfully and achieves value for money.

Direct reporting on unringfenced targeted grants

Grants with direct reporting

2.18 Departments do not have to give assurance to Parliament that local authorities have used unringfenced targeted grants for a specific activity. However, 64 per cent (£5.0 billion) of unringfenced targeted funding provided to local authorities has some form of reporting (see **Figure 5**).

Figure 5

Reporting arrangements for unringfenced targeted grants 2013-14

Most targeted funding has some form of direct reporting from the recipient

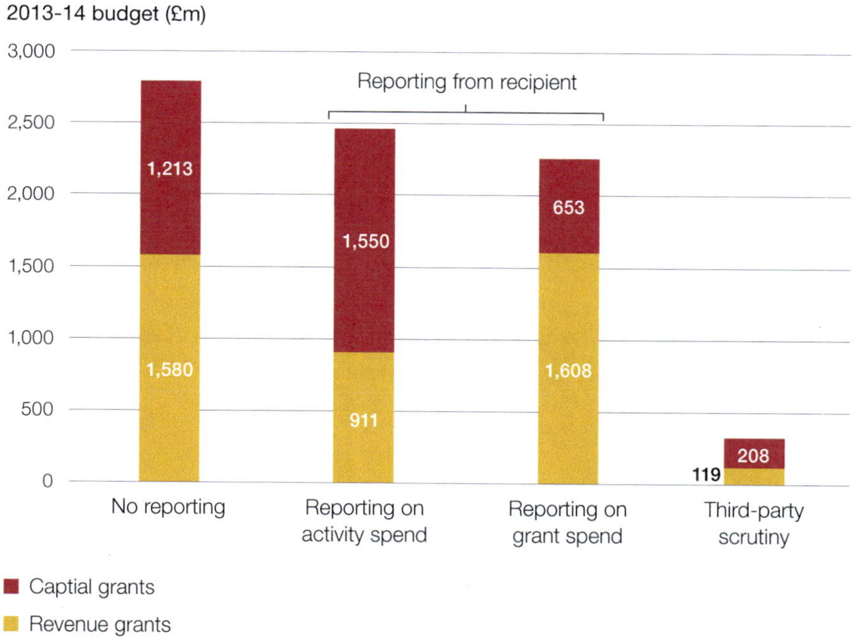

Source: National Audit Office analysis of Department for Communities and Local Government's data

2.19 Some £2.3 billion of unringfenced targeted funding involves direct reporting by the local authority, on how it has spent the grant. The PFI grant, for instance, requires local authorities to submit funding claims. These specify the purpose of expenditure and give details about the project the grant supports.

2.20 For £2.5 billion of unringfenced targeted funding, local authorities report back on how much money they spent on the relevant activity, rather than reporting on how they used the grant specifically. For instance, the Department for Transport records how local authorities use the Highways Maintenance grant through the existing Capital Outturn Return data set. In that data, local authorities record total spending on highways maintenance, rather than use of the grant alone.

2.21 The Department for Education has similar arrangements for its Education Service Grant and Basic Need funding. Local authority expenditure is captured through the Section 251 return and the School Capacity collection respectively. In both cases, local authorities record all spending on that activity, rather than only spending funded by the grant. The Section 251 return does not capture any instances where local authorities have used the grants to fund other activities. However, the School Capacity collection additionally asks local authorities to include information about how the whole grant has been spent.

Grants without direct reporting

2.22 In many cases, however, unringfenced targeted funding does not have any direct reporting arrangements. This applies to £1.6 billion of revenue funding from seven revenue grants and £1.2 billion across four capital grants.

2.23 However, our analysis shows that other methods of guiding local authority spending advocated by the Department are present. Local authorities must sign a memorandum of understanding for £859 million of this funding, £533 million is to support statutory duties, and £10 million is for the work of a third party (Local Enterprise Partnerships). This leaves only one unringfenced targeted revenue grant, the Department for Work & Pensions' Local Welfare Provision Grant (£178 million in 2013-14) where these types of factors do not apply. In this instance, the Department for Work & Pensions concluded that it would not be appropriate to ringfence the funding as local authorities needed to be able to "flex" the provision in a way that meets the needs of their local community. Nonetheless, in our view, there was a clear expectation that local authorities should use the funding for local welfare provision as set out in the settlement letter from the Secretary of State for Work and Pensions.

2.24 These same methods that guide local authorities' spending are also present for some unringfenced targeted capital grants that lack reporting mechanisms. The Department for Education provides £749 million and the Department of Health £129 million in unringfenced targeted capital grants with no reporting, for local authorities to meet their statutory duties. The Department for Transport provides a further £15 million via a bidding process. Only the Department for Transport's Integrated Transport Block (£320 million) lacks any of these additional elements and direct reporting.[23]

2.25 These additional measures mean that of the £7.8 billion in unringfenced targeted grants departments give local authorities there is:

- some formal reporting on how funding is used for £5.0 billion (64 per cent) of the funding;
- indirect methods (such as legal duties or business cases), that the Department considers influence local authorities' spending decisions for £2.3 billion (29 per cent); and
- No constraining factors or reporting for two grants worth £0.5 billion (6 per cent) in 2013-14.

Issues and risks

Absence of impact information

2.26 In general, departments do want to know how local authorities have used unringfenced targeted grants. Without direct reporting, departments do not ultimately know what impact their funding has had. Departments are aware of this issue. They are seeking to rectify gaps in their knowledge, while staying within the policy of encouraging local financial flexibility and reducing reporting burdens.

2.27 For example, the Department for Work & Pensions' system statement states there is no reporting data on the Local Welfare Provision Grant so the Accounting Officer is "unable to judge what benefits the funding has provided".[24] Consequently, the Department for Work & Pensions will review how local authorities used the funding in 2014-15. The grant will end in 2015-16 and local authorities will instead fund local welfare provision from the general grant to local authorities provided by the Department.

23 Expenditure funded by the Department for Transport's Integrated Transport Block is captured in the Capital Outturn Return in the same way as funding provided by the Highways Maintenance grant. Funding from the Integrated Transport Block is to support a diverse range of activities including tackling pollution, easing congestion and promoting accessibility. These activities are not individually identifiable within the Capital Outturn Return in the way that spending on highways maintenance is.

24 Department for Work & Pensions, *Accountability System Statement for Work and Pensions*, 2013, p.5.

2.28 The Education Funding Agency, which distributes the Department for Education's schools grants, raised concerns in 2013 over the lack of reporting on capital funding streams to local authorities. It found that it was difficult to gain assurance that local authorities have used the funding for the intended purposes, as this funding is neither time-bound nor ringfenced. The Department for Education has since started to collect data from local authorities on their total spending on the creation of new school places, with the first data collected through the School Capacity collection in summer 2013. Some of this data has been published in Basic Need scorecards, which present comparator information on output and spend on this activity by other local authorities.

2.29 The Department for Transport also knows that it has limited information on how the £320 million it gives local authorities via the Integrated Transport Block is used. The Department for Transport plans to assess how local authorities have used this funding. It is also changing how it allocates the funding to reflect local authorities' performance in achieving the scheme's objectives.

Part Three

Assurance on spending from the local accountability system

3.1 In addition to direct reporting on individual grant streams, departments secure assurance from the local system of accountability. The system is the departments' main source of assurance that local spending represents value for money.

3.2 This part of the report shows how the Department for Communities and Local Government (the Department) gets assurance over value for money through the local accountability system. It also evaluates how effectively the Department oversees this, and considers the risks of changes to funding and to service delivery models.

Value-for-money assurance

Local accountability system

3.3 The government has devolved the assurance of value for money of funding it grants to local authorities. This is based on two principles:

- local authorities' prime accountability is to their local electorate, rather than central government; and
- local councillors are best placed to decide what is value for money locally.

3.4 In the Department's view, local authorities will achieve value for money in different ways according to local priorities. Local authorities allocate resources to achieve objectives that reflect local priorities, which will not be the same in all areas. The Department should not expect to know what these priorities are, and consequently what constitutes value for money, in every local authority.

3.5 *Managing Public Money* reflects government's policy to devolve decisions over value for money. It says the quality of the assurance expected of funding to local authorities "differs from that expected of central government organisations because local authorities' prime accountability is to their electorates".[25] The Committee of Public Accounts considered this in its 2011 report *Accountability for public money*. It stated that "local accountability and reformed structures do not absolve departmental Accounting Officers of their personal responsibility to gain assurance on the way funds voted to their departments are spent".

3.6 The Committee concluded that a vital principle of accountability is that Accounting Officers must "ensur[e] that there is an appropriate framework … to provide him/her with the necessary assurances and controls".[26] In response, the government proposed that Accounting Officers who provide decentralised funding should publish a system statement that clearly shows how they will get the necessary assurances for Parliament.[27]

Department's accountability system statement

3.7 The Department's system statement describes the system for local authorities (**Figure 6** on pages 32 and 33). It sets out several methods relating to assurance on value for money including:

- **The best value duty**
  Under the Local Government Act 1999, a council must "make arrangements to secure continuous improvement in the way in which its functions are exercised, having regard to a combination of economy, efficiency and effectiveness".
- **Chartered Institute of Public Finance and Accountancy statement – on the role of local authority chief finance officers**
  This says that chief finance officers must promote and give good financial management, to safeguard public money and ensure it is used appropriately, economically, efficiently and effectively.
- **Internal and external scrutiny on spending decisions**
  Council spending decisions can be subject to public scrutiny, or by local authorities' own overview and scrutiny committees.
- **Community right to challenge**
  Created under the Localism Act 2011 and subsequent regulations, local authorities must consider and respond to expressions of interest from groups that want to take over an underperforming local service.
- **Transparency data**
  Local authorities must publish a range of data including all spending over £500.

25 HM Treasury, *Managing Public Money*, July 2013, paragraph 7.10.2.
26 HC Committee of Public Accounts, *Accountability for public money*, Session 2010-11, HC 740, April 2011.
27 Department for Communities and Local Government, *Accountability: adapting to decentralisation*, September 2011.

3.8 The Department also relies on information from external auditors of local authorities' accounts. The Audit Commission collates the results of local auditors' work on every local authority, and publishes it annually in *Auditing the Accounts*.[28] The Department uses this assessment of how local authorities are managing their funds, and local authorities' arrangements for achieving value for money. The Department is considering how to replace this assurance when the Audit Commission closes in 2015.

How the Department uses the local accountability system

3.9 The Department derives assurance on value for money of funding it gives local authorities through the effective operation of the local accountability system. Each local authority has a system of checks and balances, which influences its spending. This includes the work of regulators, scrutiny committees and elections. Effective systems assure the government that local authorities will make decisions based on local priorities that represent value for money.

3.10 This means that the Department does not have to scrutinise the value for money of local authorities' spending. Instead, it must ensure that a local system is in place and operates effectively. This is a shift from earlier arrangements. Before, the Department had access to extensive information on local authority performance. The Department, and bodies such as the Audit Commission, used the information to benchmark and compare local authorities' performance. Because of policy changes to reduce local authorities' reporting burdens and changes to comparative inspection regimes, the Department is less able to compare local authorities' performance.

3.11 Also, the Department believes that direct comparisons and benchmarking are no longer always meaningful, although it may be appropriate for certain specific statutory services, such as children's services. The increased flexibility local authorities have over their finances and their greater ability to focus on local priorities mean that the Department cannot easily compare spending between different places. Differences in spending patterns and outcomes may reflect local priorities rather than differences in performance.

Other departments' system statements

3.12 Five other departments that grant funds directly to local authorities have published accountability system statements. They rely mainly on the core local accountability system as set out in the Department's system statement. The Department says that it oversees the core system, which "can be relied upon by all other departmental accounting officers who provide funding to local authorities".[29]

28 Audit Commission, *Auditing the Accounts 2012-13*, available at: www.audit-commission.gov.uk/wp-content/uploads/2012/12/Auditing-the-Accounts-LG-2012-13.pdf

29 Department for Communities and Local Government, *Accounting Officer Accountability System Statements for Local Government and for Fire and Rescue Authorities*, September 2013, p. 7.

Throwout here

Throwout here

3.13 Other departments' system statements describe additional sources of assurance over specific local authority service areas for which they have policy responsibility. For example, the Department for Education uses Ofsted inspections (see **Figure 7**).

3.14 The Department coordinated departments' work to draft their first system statements in 2011-12. In its report *Accountability: Adapting to decentralisation* the Department gave other departments guidance on what a system statement should cover and what questions it might face when under scrutiny.

3.15 There has been limited coordination between departments since the first versions of their system statements. Updates are not coordinated and departments have updated their system statements at different times. The Department for Transport say that they are planning to update their system statement annually, bringing them into line with the Department's plans. The Department for Education says that it is currently revising its system statement. It intends to update its system statement when significant changes are needed and it does not plan to update the statement annually.

Figure 7
Departments funding local authorities that have published accountability system statements

Department	Latest iteration publication date	Local authority service coverage	Service-specific methods, as well as the Department's core system
Department for Education	September 2012	Local authority maintained schools, children's services	Role of Director of Children's Services and Lead Member for Children's Services, Ofsted inspections
Department for Environment, Food & Rural Affairs	July 2013	Flood prevention, waste services	Scrutiny of value for money and outcomes for specific projects and programmes
Department of Health	August 2012	Public Health	Publication of Public Health data against outcomes framework, director of Public Health annual report, Health and Wellbeing Boards
Department for Transport	September 2012	Local transport projects and maintenance	Highways Maintenance Efficiency Programme, approvals process for bid-based funds
Department for Work & Pensions	July 2013	Administration of Housing Benefit and local welfare provision	Data returns from local authorities to DWP

Source: National Audit Office analysis of departmental accountability system statements

Department's oversight of the local accountability system

3.16 Treasury guidance states that departments must give an annual account of how they secure assurance over how local bodies use government grants.[30] Departments should update system statements to reflect any major changes to local authority funding or the local accountability system. When departments update their system statements it acts as a check on whether the system provides the necessary assurance.

3.17 The Department has improved how it oversees the local accountability system since publishing its first system statement. It now reviews its effectiveness routinely. The Department's first version, published with its report *Accountability: Adapting to decentralisation* in 2011, did not set out how the Accounting Officer knew that it was working in practice.[31]

3.18 The Department now gives the Accounting Officer advice twice a year on the local system's effectiveness; in December and then in June, enabling highlights to be included in its annual report. Providing the Accounting Officer with regular advice encourages the Department to assess continually whether the system is working effectively.

Information to assess the local accountability system

3.19 The latest version of the Department's system statement, published in September 2013, shows what information it uses to understand whether the local accountability system is effective. This includes the following:

- Personal contacts between departmental officials and local authorities.
- Connections with organisations that write reports on the challenges facing the sector, such as the Local Government Association.
- Contacts with regulators, including the Audit Commission, Ofsted and the Care Quality Commission.
- Information shared with the Local Government Association on its sector-led improvement activity.
- Publicly available data generated by local authorities on a range of council services.
- Direct contact from local authorities, such as requests for permission to treat revenue costs as capital costs (capitalisation requests).[32]

30 HM Treasury, *Managing Public Money*, July 2013, paragraph 7.10.3.
31 Department for Communities and Local Government, *Accountability: adapting to decentralisation*, September 2011.
32 Department for Communities and Local Government, *Accounting Officer Accountability System Statements for Local Government and for Fire and Rescue Authorities*, September 2013.

3.20 The Department uses this information to "understand whether the system requires any changes, and would have the information to know whether intervention by the Department was needed".[33] Using this information, the Department is satisfied that the local accountability system is effective, which allows the Accounting Officer to give Parliament the necessary assurances. The Department advised the Accounting Officer in December 2013 that the system did not require changes.

3.21 The Department has taken steps to assess whether the local accountability system is effective, and it uses the system to monitor the sector. However, it is not clear that the Department really knows whether the system is effective in securing value for money. The information it uses to monitor the system gives limited insight into whether the system is effective in securing value for money. It focuses more strongly instead on financial and service sustainability. For instance:

- Contacts between departmental officials and local authorities are normally to communicate policy decisions and receive feedback.
- Other organisations that publish reports, which the Department uses, focus mostly on whether local authorities are financially sustainable.
- The Department does not actively use local authorities' performance and financial data to oversee the system's effectiveness – though this information is available for most local authorities.
- The Department accepts that capitalisation requests mostly indicate financial stress rather than value for money concerns.

3.22 Consequently, the Department has some understanding of the risk of financial and service failure in the local authority sector. However, it could still better understand whether the system is effective in securing value for money. The Department's advice to the Accounting Officer on the system statement did not consider the effectiveness of the local accountability system towards achieving value for money. The Department did not assess whether methods that are directly responsible for achieving value for money locally were effective. The focus of the update is instead on local authorities' financial and service sustainability.

33 Department for Communities and Local Government, *Accounting Officer Accountability System Statements for Local Government and for Fire and Rescue Authorities*, September 2013, p.20.

Issues and risks of the Department's approach

3.23 Generally, the system that the Department sets out is coherent and consistent. However, there are two potential areas of concern.

Unringfenced targeted grants

3.24 The Department relies on local accountability, which works for unringfenced general grants but is less effective for unringfenced targeted grants.

3.25 In the Department's view, unringfenced targeted grants allow national priorities to be pursued locally while also providing a degree of local financial flexibility that ringfenced funding would not. Local authorities can reallocate unspent funding from these types of grants to other activities rather than repaying it to departments. The Department also believes that allocating precise, ringfenced budgets to individual services can lead to poor value for money. This could lead local authorities to use all of the allocation on objectives defined by government regardless of whether they reflect local priorities. Additionally, the Department does not consider there to be accurate information on the local costs of providing services in a given area to set ringfenced funding at the optimum level.

3.26 However, the combination of unringfenced targeted grants and an approach to value for money in which local needs and objectives are prioritised, creates a potential assurance gap. With these grants, departments exert some influence over local authorities' decisions by setting spending expectations rather than leaving the decision solely for local consideration. While a department may expect local authorities to use this funding in particular ways, local decisions may override the Department's expectations. This is reinforced by the reduction in direct reporting on these types of grants, which means that departments may not know whether local authorities have met their objectives.

3.27 We have previously reported on cases where departments could not assess value for money of unringfenced targeted grant funding. In *Reducing costs in the Department for Transport* we reported that the Department for Transport held very limited information on value for money of funding for highways maintenance. This was because local authorities did not have to spend the Department's funding on maintenance or give an account to the Department for it.[34]

34 Comptroller and Auditor General, *Reducing costs in the Department for Transport*, Session 2010-12, HC 1700, National Audit Office, December 2011.

3.28 Similarly, our report *Capital funding for new school places* found the Department for Education had limited feedback to assure itself that overall the system for new school places was achieving value for money.[35] It has since taken steps to address this through the collection of data about local authorities' spending on new school places and introduced Basic Need scorecards, including publication of comparative information.

3.29 A system in which departments want to achieve specific objectives by funding local authorities is inconsistent with an assurance and accountability framework where local priorities take precedence, and where there is limited reporting on local spending. However, our analysis of changes in types of grants suggests that the level of unringfenced targeted grants is falling, as the new system beds in.

Managing future risks to assurance

3.30 The local accountability system in the Department's system statement relies on local councillors being democratically accountable to the local electorate and appointed officers having legal responsibilities. These elements of the system are specific to individual local authorities, which creates difficulties where spending activity is cross-border or multi-agency.

3.31 Our 2013 study *Funding and structures for local economic growth* showed these issues within Local Enterprise Partnerships.[36] Each partnership covers a number of local authority areas. The Department uses the accountability system statement to assure Parliament on Local Enterprise Partnerships' spending. Local business leaders and civic leaders run the partnerships, with one local authority overseeing how government funding is distributed and used for each. One local authority could therefore allocate funds for other local authorities, subject to the parameters of a partnership agreement. This would blur the lines of accountability between local authorities and their electorates, on which the local accountability system hinges. Local authority officers' statutory duties also become more complex, as they apply only to a specific local authority. Though they retain responsibility for allocating funds to the Partnership, they may not be involved ultimately in decisions over how it is spent.

35 Comptroller and Auditor General, *Capital funding for new school places*, Session 2012-13, HC 1042, National Audit Office, March 2013.

36 Comptroller and Auditor General, *Funding and structures for local economic growth*, Session 2013-14, HC 542, National Audit Office, December 2013.

3.32 The Department also needs to consider the implications of greater levels of multi-agency working within local authority areas. Increasingly, government is encouraging local authorities to work with other local public bodies in sectors such as health, policing and welfare to maximise the local impact of public spending via pooled budgets, and to tackle complex, multi-faceted local issues. Health and Wellbeing Boards and the introduction of pooled social care budgets are primary examples. However, local agencies remain accountable to individual departments via their separate accountability systems, meaning there is misalignment of local bodies' joint delivery with their discrete lines of departmental accountability. Departments' accountability systems need to complement one another in order to provide a sufficient degree of assurance to departments, and therefore Parliament, on jointly delivered local schemes without undermining local integration and joint working.

3.33 The Department recognises the potential challenge that multi-agency delivery models might present to traditional accountability systems. The Department has been meeting with representatives from local government and other local service providers together with the relevant government departments to explore how accountability systems could better support service transformation.

Glossary

Accountability system statement	Departments' Accounting Officers draft these statements, which explain how they achieve accountability for the grants they distribute to local bodies. The statements describe systems of checks and balances, which provide assurances that local bodies spend their resources with regularity and value for money.
Ambit	Each department's ambit sets out the list of activities upon which it can generate income and incur expenditure. Each department's ambit needs to be approved by Parliament.
Capital funding	Funding for investment in non-current assets such as land, buildings, plant and machinery.
Estimates process	The means by which government obtains authority from Parliament to use funds to finance departments' agreed spending programmes for each financial year.
Grant certification	Auditors certify claims and returns to provide assurance to grant-paying bodies that claims for grants and subsidies comply with terms and conditions, or that information in financial returns is reliable.
Grant determination	Grant-giving departments give determinations to local authorities to explain the amount and purpose of each grant.
Housing Revenue Account	Where local authorities account for all expenditure incurred and income received in respect of the provision of local authority owned rented accommodation.
National Indicator Set	Effective from April 2008 to March 2011, this was the set of indicators, which central government used to performance-manage local government.
Prudential borrowing	Unsupported borrowing funded by local authorities' own resources, such as Council Tax, rather than through government grants.
Revenue funding	Funding for the cost of running local authority services within each financial year. This includes the costs of staffing, heating, lighting and cleaning, together with expenditure on goods and services consumed within the year.

Regularity	The principle that public resources are used in accordance with Parliament's authority, which comes through legislation, through the Estimate process and through the authority which Parliament delegates to HM Treasury.
Ringfenced grants	Local authorities must use these grants for a specific purpose. Departments can claw back funding if local authorities do not use it, or do not spend it in line with funding conditions.
Single Data List	The list of all the data sets that local authorities must submit to central government. This replaced the National Indicator Set and commenced in April 2011.
Section 151 officer	Each local authority has a nominated officer responsible for making arrangements for the proper administration of its financial affairs. The Section 151 officer is normally the local authority's treasurer.
Unringfenced general grants	Local authorities can use these grants for any purpose within their legal powers.
Unringfenced targeted grants	Government departments intend for local authorities to use these grants for a specific purpose. Nonetheless local authorities can reallocate funding to other activities should they choose, and departments typically have no power to claw back this funding.

Appendix One

Our audit approach

1 This report examines how the government funds local authorities, the conditions attached to this funding, and how it gives Parliament assurance on how local authorities use that funding. It focuses on the role of the Department for Communities and Local Government, which leads central government responsibility for this area.

2 We reviewed:

- how the government has changed how it funds local authorities since 2010-11 and the Department's role in implementing these changes;
- the local authorities' direct reporting arrangements for the grants they receive, and how far they give departments enough assurance on regularity, and information on the impact of departmental spending; and
- how effectively the government manages its arrangements for assuring local government funding to provide necessary assurances to Parliament.

3 Our audit approach is summarised in **Figure 8**. Our evidence is described in Appendix Two.

Figure 8
Our audit approach

The Department's objective

The main objectives were to:

- increase local authorities' financial flexibility;
- reduce reporting burdens on local authorities; and
- assure Parliament that local authorities spend the money as Parliament intended and achieve value for money locally.

How this will be achieved

The government has removed ringfences that say how local authorities should spend their funding from most government grants. They have reduced the amount of direct reporting on grants. Departments rely on local systems of checks and balances, as described in accountability system statements to get the right assurance for Parliament.

Our study

We assessed the Department's role in changing local authorities' funding and whether they, along with other departments that fund local authorities, are overseeing funding arrangements that give the necessary assurances to Parliament for the funding.

Our evaluative criteria

The Department led government in changing how it funds local authorities in line with the intentions of the 2010 spending review.

The Department ensured that local authorities' funding across government was consistent and coherent, with assurance arrangements that matched the condition of grants, and gave departments enough information on policy impact.

The Department is effectively overseeing the assurance arrangements for local authorities, so departments give Parliament the right assurances.

Our evidence

(see Appendix Two for details)

We reviewed the Department's role in changing how government funds local authorities, by:

- interviewing departmental officials and reviewing documents; and
- analysing local authority financial data.

We assessed the direct reporting arrangements in place for funding granted to local authorities, by:

- interviewing departmental officials;
- reviewing departments' grant documentation, that detail grant intentions and conditions; and
- reviewing evidence underpinning the audit opinion over the regularity of departmental expenditure.

We assessed how effectively the Department oversees the assurance arrangements, by:

- interviewing departmental officials;
- reviewing department documents, including accountability system statements; and
- drawing on findings of our previous reports that assess the value for money of funding to local authorities.

Our conclusions

The Department has implemented the government policies to increase local authorities' flexibility over their funding and reduce reporting burdens. Local authorities have more freedom to allocate resources according to local priorities. However, at the same time reductions to government funding have put increasing pressure on budgets. The Department's arrangements to assure Parliament over funding are in transition as government adapts to receiving less evidence on how local authorities spend funding. It is moving to a model of local accountability focused on understanding, rather than shaping, how local authorities use funding. Tensions remain where departments continue to specify policy objectives for grants, despite local priorities now taking precedence for how funding can be spent.

Under the new arrangements the Department does not have to monitor local spending decisions. However, it still must assure itself that the local accountability system that oversees local spending is effective. The Department has improved its understanding of how the system prevents financial and service failure. It could do more to understand whether the system is effective in delivering value for money.

Appendix Two

Our evidence base

1 We collected the evidence below between February and April 2014. We examined how the Department led government's changes to local authority funding and oversees arrangements that give Parliament the right assurances.

2 We used quantitative and qualitative techniques in our examination. Our audit approach is outlined in Appendix One.

Changes to government funding (Part One)

3 We reviewed legislative documents that set out the government's intentions to increase local flexibility, including the 2010 spending review and the Localism Act.

4 We analysed local authority Revenue Outturn, Revenue Account and Capital Outturn data to identify the sources of local authority income. We obtained budgeted grant amounts from the local government finance settlements for 2010-11 and 2013-14. The Department also gave detailed information on revenue grants that it provided in 2010-11.

5 We reviewed departmental documents to understand the conditions on grants that the departments pay to local authorities – including legislation, grant guidance and explanatory memorandums. We reviewed documents that show how the Department coordinated the changes to local authority funding across government after the 2010 spending review, including guidance documents to other departments.

6 We interviewed Department officials to understand how they made the changes to local authority funding. We also interviewed officials from HM Treasury, the Department for Education and the Department for Transport to understand their role in the process to change the conditions on local authority funding.

Direct reporting arrangements for local grants (Part Two)

7 We assessed how far local authorities' reporting their use of grants gives sufficient assurance on regularity, and information on the impact of departmental spending.

8 We have reviewed department documents that set out for each grant how the grant-giving departments assure that local authorities have spent funding in line with Parliament's intentions. We assessed the fit between the grant type and conditions for each grant and the nature of any associated direct reporting.

9 We interviewed officials from the Department, the Department for Education and the Department for Transport to see how they make assurance systems that are appropriate to grant conditions, focusing on unringfenced targeted grants.

10 We spoke with the Local Government Association and the Audit Commission to understand their views on changes in arrangements for direct reporting.

11 We have worked with our internal teams responsible for auditing relevant departments to understand their methods for getting assurance on regularity.

12 We have drawn on our value for money studies that examine different aspects of assurance for local authority funding.

Assurance over value for money (Part Three)

13 We examined how the Department gets assurance on how central government funds local government.

14 We reviewed department documents, including their accountability system statements, and the Department's advice to the Accounting Officer on the system's effectiveness.

15 We interviewed officials from the Department, the Department for Education and the Department for Transport to understand how they oversee the local accountability system to ensure that it is effective.

16 We spoke with the Local Government Association and the Audit Commission to understand their views on changes to the assurance on value for money on local spending.

Appendix Three

Government grants paid to local authorities, 2013-14

Figure 9
Classification of grants paid to local authorities in 2013-14[1]

Grant name	Amount (£m)	Department	Revenue or capital	Grant type	Reporting mechanism
Public Health Grant	2,660	DH	Revenue	Ringfenced	Reporting on grant spend
Adoption Reform Grant[2]	150	DfE	Revenue	Ringfenced, targeted[1]	Reporting on grant spend
Asylum Seeking Children Grant	65[3]	HO	Revenue	Ringfenced	Reporting on grant spend
Air Quality Management	1	DEFRA	Capital	Ringfenced	Reporting on grant spend
Court Desk Revenue Grant	1	DCLG	Revenue	Ringfenced	Reporting on grant spend
Phonics Screening Check	0.1	DfE	Revenue	Ringfenced	Reporting on grant spend
Local Major Transport Schemes	456	DfT	Capital	Ringfenced	Independent scrutiny
KS2 Monitoring and Moderation	3	DfE	Revenue	Ringfenced	Independent scrutiny
Right to Control Trailblazers	2	DWP/DCLG/DH	Revenue	Ringfenced	Independent scrutiny
Demonstration Projects	1	DCLG	Revenue	Ringfenced	Independent scrutiny
Revenue Support Grant	14,483	DCLG	Revenue	Unringfenced general	No reporting
Retained income from Rate Retention Scheme	9,511	DCLG	Revenue	Unringfenced general	No reporting
New Homes Bonus	668	DCLG	Revenue	Unringfenced general	No reporting
Council Tax Freeze Grant 2013-14	265	DCLG	Revenue	Unringfenced general	No reporting
New Homes Bonus Topslice	82	DCLG	Revenue	Unringfenced general	No reporting
Rural Area Efficiency Grant	9	DCLG	Revenue	Unringfenced general	No reporting
Funding transfer from the NHS to social care	859	DH	Revenue	Unringfenced targeted	No reporting
Capital Maintenance	749	DfE	Capital	Unringfenced targeted	No reporting
Housing Benefit Subsidy Admin Grant	402	DWP	Revenue	Unringfenced targeted	No reporting

Figure 9 *continued*

Classification of grants paid to local authorities in 2013-14[1]

Grant name	Amount (£m)	Department	Revenue or capital	Grant type	Reporting mechanism
Integrated Transport Block	320	DfT	Capital	Unringfenced targeted	No reporting
Local Welfare Provision Grant	178	DWP	Revenue	Unringfenced targeted	No reporting
Department of Health Community Capital Grant	129	DH	Capital	Unringfenced targeted	No reporting
Local services support grant	56	DCLG	Revenue	Unringfenced targeted	No reporting
Local Reform and Community Voices	42	DH	Revenue	Unringfenced targeted	No reporting
Council Tax support New Burdens Funding	33	DCLG	Revenue	Unringfenced targeted	No reporting
Community Linking Places Fund	15	DfT	Capital	Unringfenced targeted	No reporting
Local Enterprise Partnerships Core Funding	10	DCLG	Revenue	Unringfenced targeted	No reporting
Basic Need	800	DfE	Capital	Unringfenced targeted	Reporting on activity spend
Education Services Grant	799	DfE	Revenue	Unringfenced targeted	Reporting on activity spend
Highways Maintenance block grant	750	DfT	Capital	Unringfenced targeted	Reporting on activity spend
Weekly Collection Support Scheme	112	DCLG	Revenue	Unringfenced targeted	Reporting on activity spend
Private Finance Initiative grant	1,219	DCLG/DfE	Revenue	Unringfenced targeted	Reporting on grant spend
Decent Homes Backlog Funding	195	DCLG	Capital	Unringfenced targeted	Reporting on grant spend
Disabled Facilities Grant	180	DCLG/DH	Capital	Unringfenced targeted	Reporting on grant spend
Troubled Families Grants	149	DCLG	Revenue	Unringfenced targeted	Reporting on grant spend
Additional Highways Maintenance Funding	140	DfT	Capital	Unringfenced targeted	Reporting on grant spend
Pinch Point Fund	70	DfT	Capital	Unringfenced targeted	Reporting on grant spend

Figure 9 *continued*

Classification of grants paid to local authorities in 2013-14[1]

Grant name	Amount (£m)	Department	Revenue or capital	Grant type	Reporting mechanism
Council tax support transitional grant	45	DCLG	Revenue	Unringfenced targeted	Reporting on grant spend
Short Breaks	40	DfE	Capital	Unringfenced targeted	Reporting on grant spend
Regional Growth Fund	23[3]	BIS/DCLG	Revenue/Capital	Unringfenced targeted	Reporting on grant spend
Coastal Communities Fund	22	DCLG	Revenue	Unringfenced targeted	Reporting on grant spend
Preventing Repossessions Fund	19	DCLG	Revenue	Unringfenced targeted	Reporting on grant spend
Cycling Ambition	17	DfT	Capital	Unringfenced targeted	Reporting on grant spend
Housing Benefit Reform Transitional Funding	15	DWP	Revenue	Unringfenced targeted	Reporting on grant spend
Innovation Challenge Award	9	DCLG	Revenue	Unringfenced targeted	Reporting on grant spend
London Highways Maintenance	7	DfT	Capital	Unringfenced targeted	Reporting on grant spend
Community Right to Challenge	3	DCLG	Revenue	Unringfenced targeted	Reporting on grant spend
Community Right to Bid	3	DCLG	Revenue	Unringfenced targeted	Reporting on grant spend
Plugged in Places Grants	3	DfT	Capital	Unringfenced targeted	Reporting on grant spend
Contaminated Land	2[3]	DEFRA	Capital	Unringfenced targeted	Reporting on grant spend
Alcohol fund for problem drinking	0.5	DCLG	Revenue	Unringfenced targeted	Reporting on grant spend
Local Sustainable Transport Fund	92	DfT	Revenue	Unringfenced targeted	Scrutiny from grant funding body
Local Sustainable Transport Fund	84	DfT	Capital	Unringfenced targeted	Scrutiny from grant funding body

Figure 9 *continued*

Classification of grants paid to local authorities in 2013-14[1]

Grant name	Amount (£m)	Department	Revenue or capital	Grant type	Reporting mechanism
Flood Management	72	DEFRA	Capital	Unringfenced targeted	Scrutiny from grant funding body
Efficiency Support Grant	9	DCLG	Revenue	Unringfenced targeted	Scrutiny from grant funding body
Youth Contract – Core Cities	3	DfE	Revenue	Unringfenced targeted	Scrutiny from grant funding body
Clusters of Empty Homes	25	DCLG	Capital	Unringfenced targeted	Independent scrutiny
Traveller Pitch Funding	19	DCLG	Capital	Unringfenced targeted	Independent scrutiny
Neighbourhood Planning Front Runners	15	DCLG	Revenue	Unringfenced targeted	Independent scrutiny
Homelessness Change Programme	8	DCLG	Capital	Unringfenced targeted	Independent scrutiny
Household Reward and Recognition Fund	0.1[3]	DEFRA	Revenue	Unringfenced targeted	Independent scrutiny

Notes

1 Unless otherwise stated all figures quoted in the table are taken from budget plans for 2013-14. Actual grant outturn payments may vary compared to budget plans.

2 £50 million of the adoption reform grant was ringfenced, the remaining £100 million was via an unringfenced targeted grant.

3 Outturn data.

4 BIS = Department for Business, Innovation & Skills, DCLG = Department for Communities and Local Government, DEFRA = Department for Environment, Food & Rural Affairs, DfE = Department for Education, DfT = Department for Transport, DH = Department of Health, DWP = Department for Work & Pensions, HO = Home office.

Source: National Audit Office analysis of data from the Department for Communities and Local Government, the Department for Education, the Department for Environment, Food & Rural Affairs, the Department for Business, Innovation & Skills and the Home Office

Made in the USA
Monee, IL
05 November 2026

67383067R00033